Leaping Learners
Education, LLC

For more information and resources visit us at:
www.leapinglearnersed.com

Published by Leaping Learners Education, LLC, Verona, NJ 07044.

Every attempt has been made to credit each photo. Please contact us if there has been an error and we will resolve the issue.

Photo Credits

Monkey graphics © dreamblack46/stock.adobe.com

Cover, © Natalia Kuzmina/shutterstock.com; Title Pg. Ivan Kuzmin/shutterstock.com; Pg. 1-2 © Natalia Kuzmina/shutterstock.com; Pg. 3-4 © max_776/stock.adobe.com; Pg. 5-6 © Ondrej Prosicky/shutterstock.com; Pg. 7-8 © Anton_Ivanov/shutterstock.com; Pg. 9-10 © fivespots/shutterstock.com; Pg. 11-12 © dossyl/shutterstock.com; Pg. 13-14 © Shark_749/shutterstock.com; Pg. 15-16 © jeep2499/shutterstock.com; Pg. 17-18 © fivespots/shutterstock.com; Pg. 19-20 © PUSCAU DANIEL/shutterstock.com; Pg. 21-22 © Tanguy de Saint-Cyr/shutterstock.com, © Independent birds/shutterstock.com, © Pedro A. Estevez Calzado/shutterstock.com, © Elona Hart/shutterstock.com; 23-24 © Nutthagreth Chaiwong/shutterstock.com, © Oleg Blazhyievskyi/shutterstock.com; Pg. 26 © hagit berkovich/shutterstock.com; Pg. 27 © nikiteevkonstantin/shutterstock.com, © Zoart Studio/shutterstock.com

All designs by Sean Bulger

ISBN
978-1-948569-18-7

Dear Parents and Guardians,

Thank you for purchasing a *Matt Learns About* series book! After teaching students from kindergarten to second grade for more than seven years, I became frustrated by the lack of engaging books my students could read independently. To help my students engage with nonfiction topics, my wife and I decided to write nonfiction books for children. We hope to inspire young children to learn about the natural world.

Here at Leaping Learners Education, LLC, we have three main goals:

1. Spark young readers' curiosity about the natural world
2. Develop critical independent reading skills at an early age
3. Develop reading comprehension skills before and after reading

We hope your child enjoys learning with Matt. If you or your children are interested in a topic we have not written about yet, send us an email with your topic, and maybe your book will be next!

Thank you,

Sean Bulger, Ed.M

www.leapinglearnersed.com

Reading Suggestions:

Before reading this book, encourage your children to do a "picture walk," where they skim through the book and look at the pictures to help them think about what they already know about the topic. Thinking about what they already know helps children get excited about learning more facts and begin reading with some confidence.

Preview any new vocabulary words with your child. Key vocabulary words are found on the last few pages of the book. Have your children use each new phrase in their own words to see if they understand the definition.

After previewing the book, encourage your child to read the book independently more than once. After they have read it, ask them specific questions related to the information in the book. Encourage them to go back and reread the relevant section in the book to retrieve the answer in case they forgot the facts.

Finally, see if your child can complete the reading comprehension exercises at the end of the book to strengthen their understanding of the topic!

This book is best for ages 6 - 8
but. . .
Please be mindful that reading levels are a guide and vary depending on a child's skills and needs.

Matt Learns About . . . Boa Constrictors

Written by Sean and Anicia Bulger

Table of Contents

Hi! My name is Matt. I love to discover and learn new things. In this book, we will learn about boa constrictors. Let's go!

Introduction

Hiiiiissssss

The boa constrictor is a deadly snake.

Habitat

Boa constrictors live in Costa Rica and the Amazon rainforest.

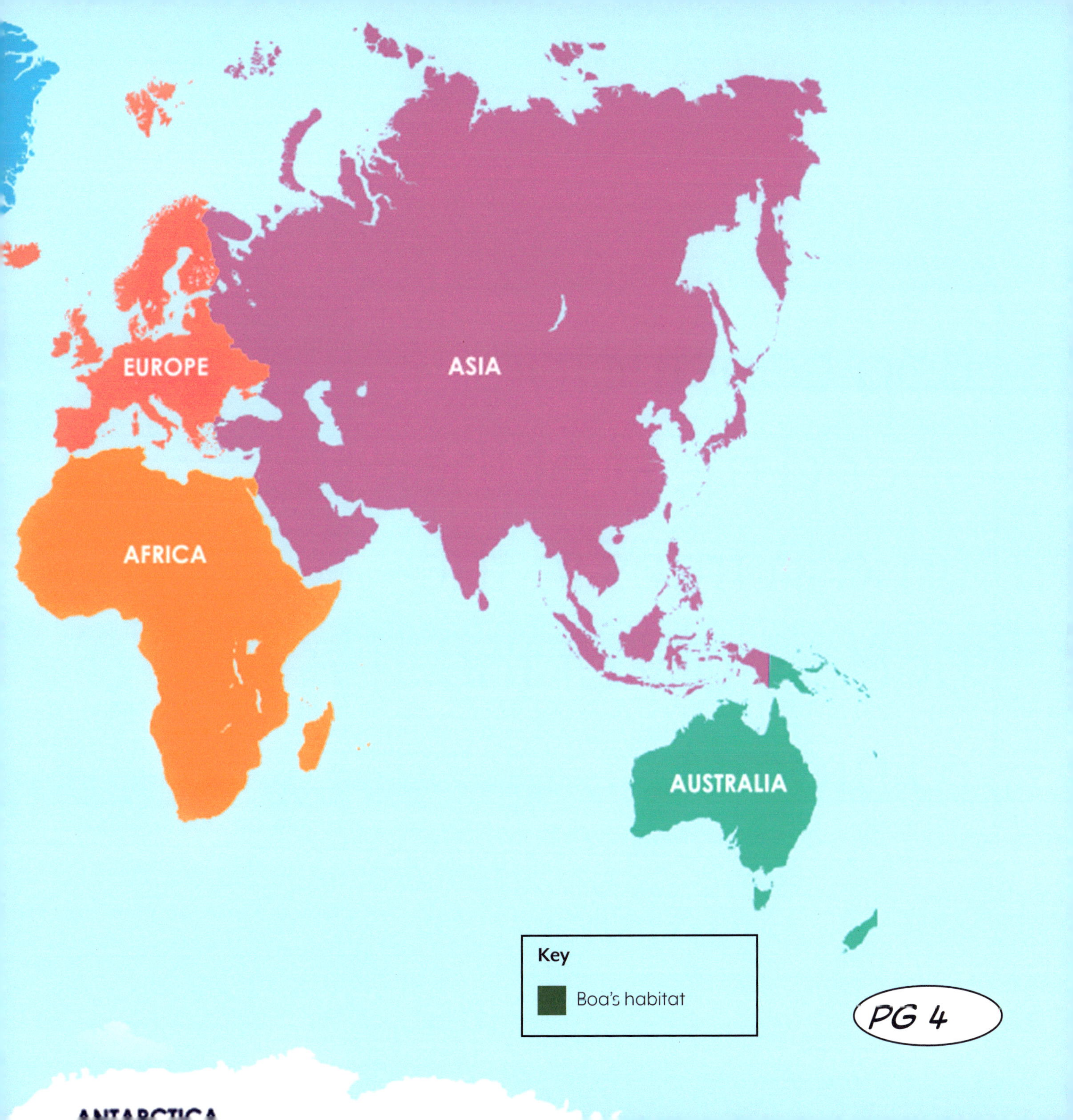
EUROPE
ASIA
AFRICA
AUSTRALIA
ANTARCTICA
Key
Boa's habitat
PG 4

Boa constrictors spend most of their time in trees in the understory of the rainforest.

Boa constrictors also live on the forest floor. They hunt for food on the forest floor.

Body

Boa constrictors are reptiles. They have brown **scales**. This helps them blend into the forest and move smoothly.

Reptiles

*Have scales

*Breathe air

*Have bones

Scales

Boa constrictors can be about six feet long. This is as tall as some adult human males.

Scales

Boa constrictors can be about six feet long. This is as tall as some adult human males.

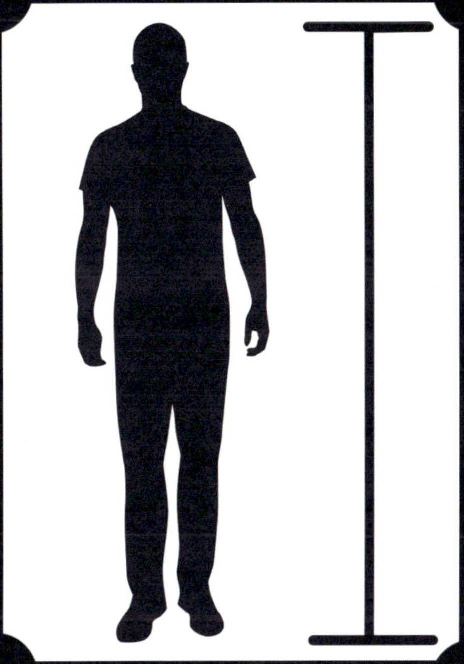

PG 12

When boa constrictors grow, they **shed** their skin and grow new skin.

PG 14

While it looks like snakes do not have bones, they actually have bones throughout their entire body.

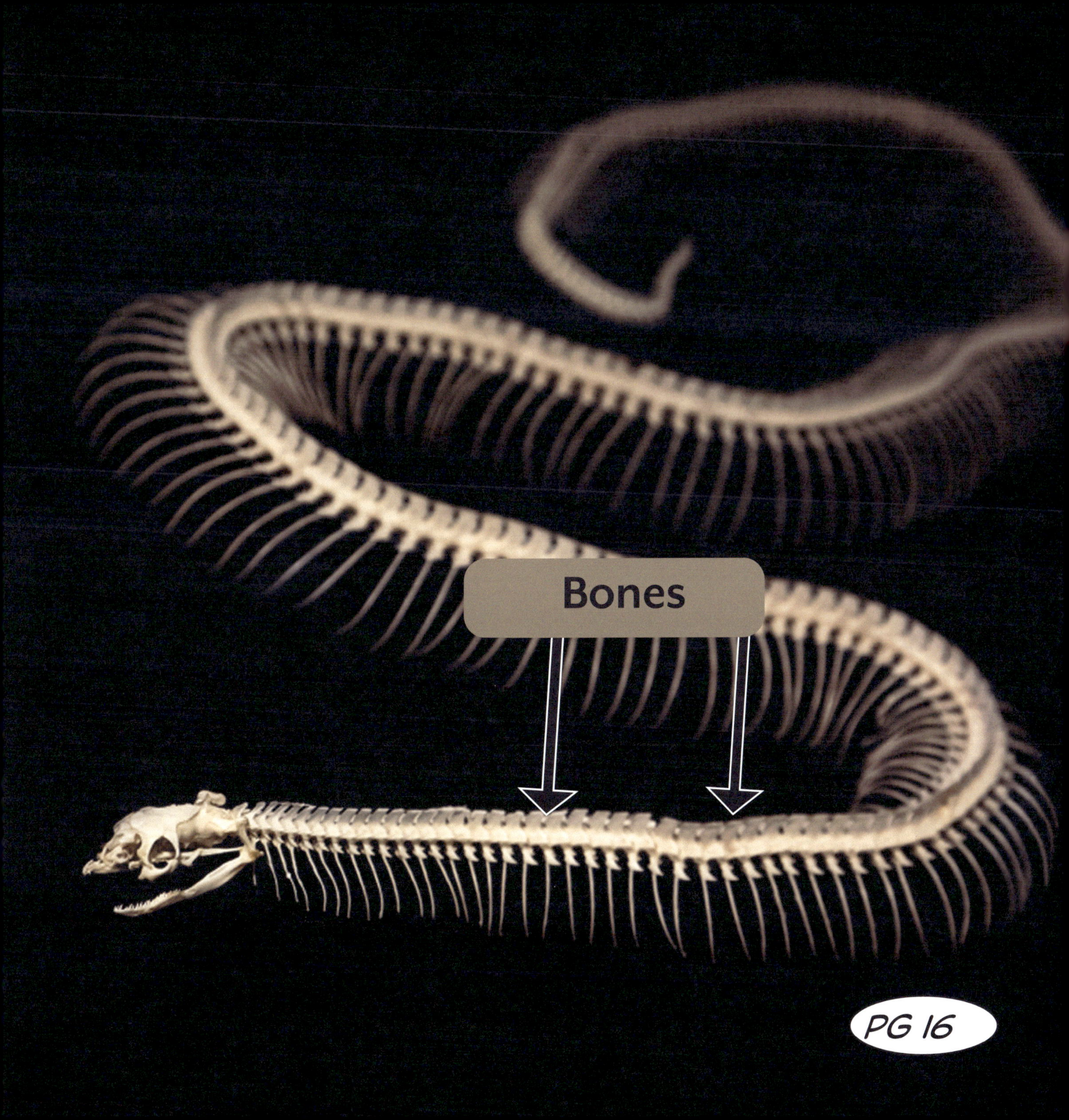
Bones

Boa constrictors have sharp **fangs** in their mouth that they use to grab their **prey**.

Fangs

Boa constrictors use their eyes and tongues to hunt for food. They smell their prey with their tongue and have very good eyesight to see their prey.

Eye
Tongue

Food

Boa constrictor's favorite foods are squirrels, rats, and bats. They **ambush** their prey from tree branches or the ground.

squirrel

Rat

Bat

What do Boa constrictors eat?

Unlike some other snakes, boa constrictors are not **venomous**.

To kill their prey, they squeeze it to death. Once the prey is dead, boa constrictors will swallow it whole.

Baby Boa Constrictors

Female boa constrictors do not lay eggs. They give birth to live young. They have about 25 babies at a time.

Once the baby is born, it is totally independent and does not need the mother to take care of it.

PG 26

The longest boa ever recorded was 18 feet long!

After a boa constrictor sheds its skin, it is partially blind for a few hours.

A boa constrictor can weigh 100 pounds!

Glossary

A glossary tells the reader the meaning of important new words.

Shed – When old skin falls off to make room for new skin

Scales – Hard covering over a reptile's skin

Fangs – Sharp and pointy teeth that a snake uses to catch prey

Prey – Animal that is food for another animal

Ambush – To sneak up on prey without the other animal knowing

Venomous – To inject poisonous liquid into other animals to kill or paralyze them

Draw a picture of a boa constrictor.

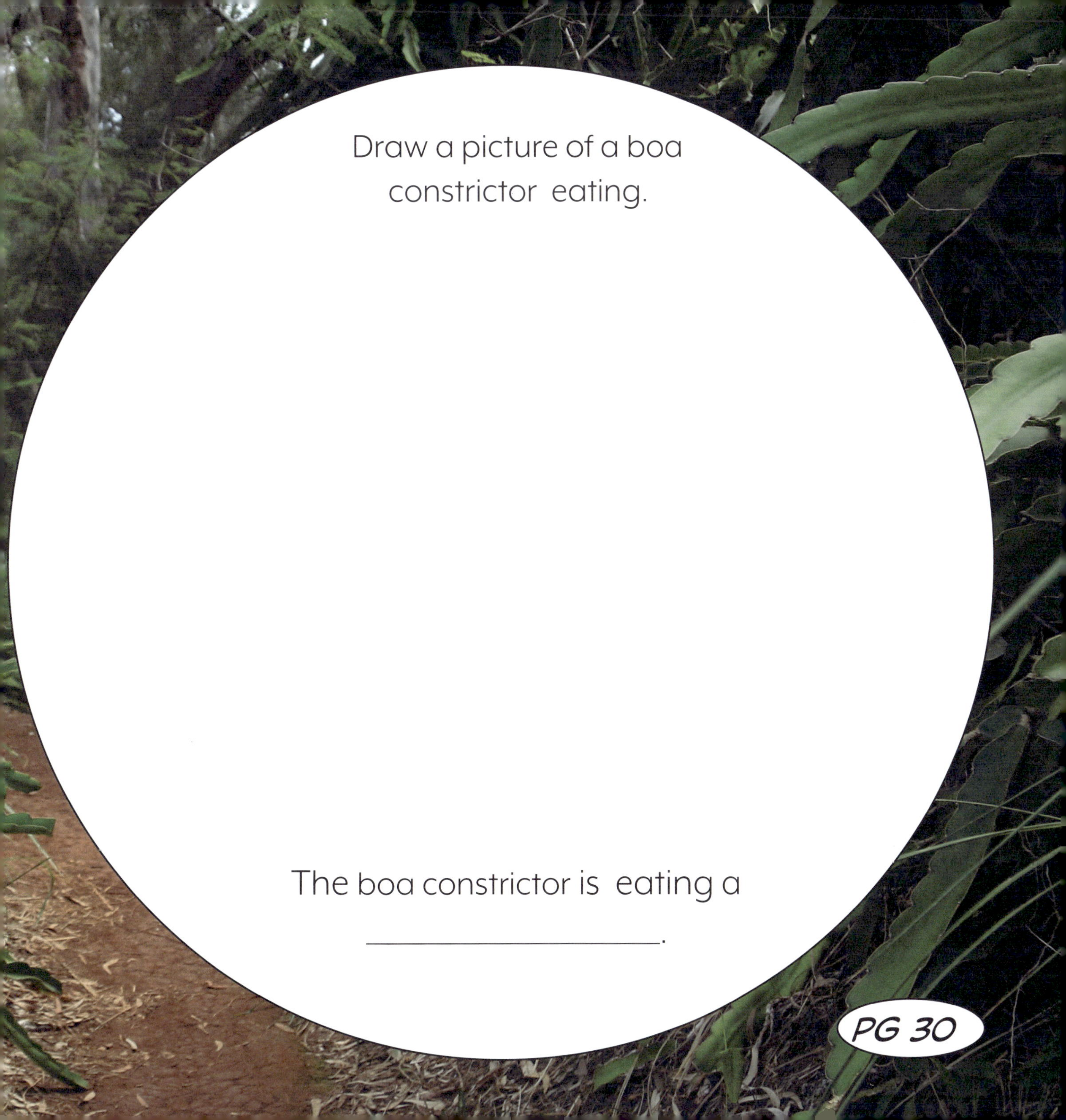

Draw a picture of a boa constrictor eating.

The boa constrictor is eating a

____________________.

Quiz

1. What is one layer of the rainforest that boa constrictors live in?

a. Canopy

b. Emergent

c. Understory

2. Why do boa constrictors have brown scales?

a. To blend into the trees

b. To scare animals away

c. To communicate with other boa constrictors

3. What is the main idea of the section called "Food"?

a. How people eat boa constrictors

b. What boa constrictors eat

c. Foods in the rainforest

Common core standards:
RI. 1. 1 - Questions 1, 2
RI. 1. 2 - Question 3

4. What is one food that boa constrictors eat?

a. Fruit

b. Rats

c. Fish

5. How do snakes smell?

a. With their nose

b. With their tongue

c. With their scales

6. What does the picture on page 24 teach you?

a. How a boa constrictor kills its prey

b. Where boa constrictors live

c. What a boa constrictor looks like

Common core standards:
RI. 1. 1 - Questions 4, 5
RI. 1.8 - Question 6

Want to learn about ocean animals? Check out the "Fay Learns About..." series!

Great for emerging readers ages 6-8

Want to learn about Farm Animals? Check out the "Katie Teaches you About..." Series!

Great for early readers ages 4-6

Want to learn about colors? Check out the "Clayton Teaches you About..." series!

Great for early readers ages 4-6

Made in the USA
Las Vegas, NV
09 November 2026

57757234R00026